For all the future genomic scientists.

**Yay Yay DNA!
Do You Wonder What Makes You You?
With Mendel G. Cat**

Art and words by Mark A. Hicks

Copyright 2025 Mark A. Hicks. All rights reserved.

No part of this publication may be reproduced, distributed, or transmitted in any form or by any means, including photocopying, recording, AI, or any other electronic or mechanical methods for any use without prior written permission. You can contact the author at www.genetionary.org/contact

ISBN 979-8-218-77999-3

A special thank you to Michelle Springer, a Certified Genetic Counselor and educator for her help and support in creating this book.

A thin thread connects all life

Do you ever wonder what makes cats cats and you you?

It's because of something called **DNA**.

YAY YAY DNA!

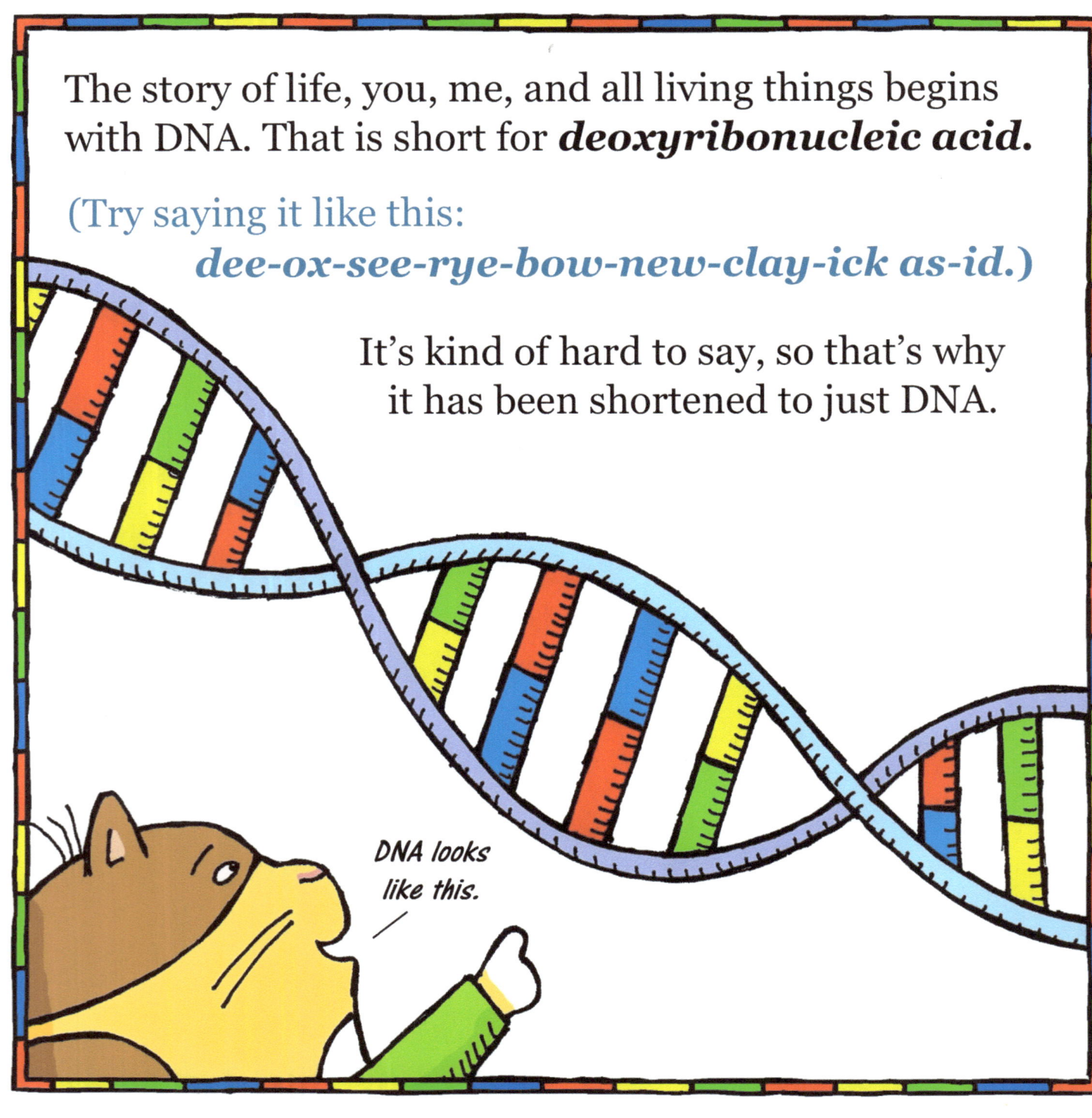

The story of life, you, me, and all living things begins with DNA. That is short for **deoxyribonucleic acid.**

(Try saying it like this:
dee-ox-see-rye-bow-new-clay-ick as-id.)

It's kind of hard to say, so that's why it has been shortened to just DNA.

DNA looks like this.

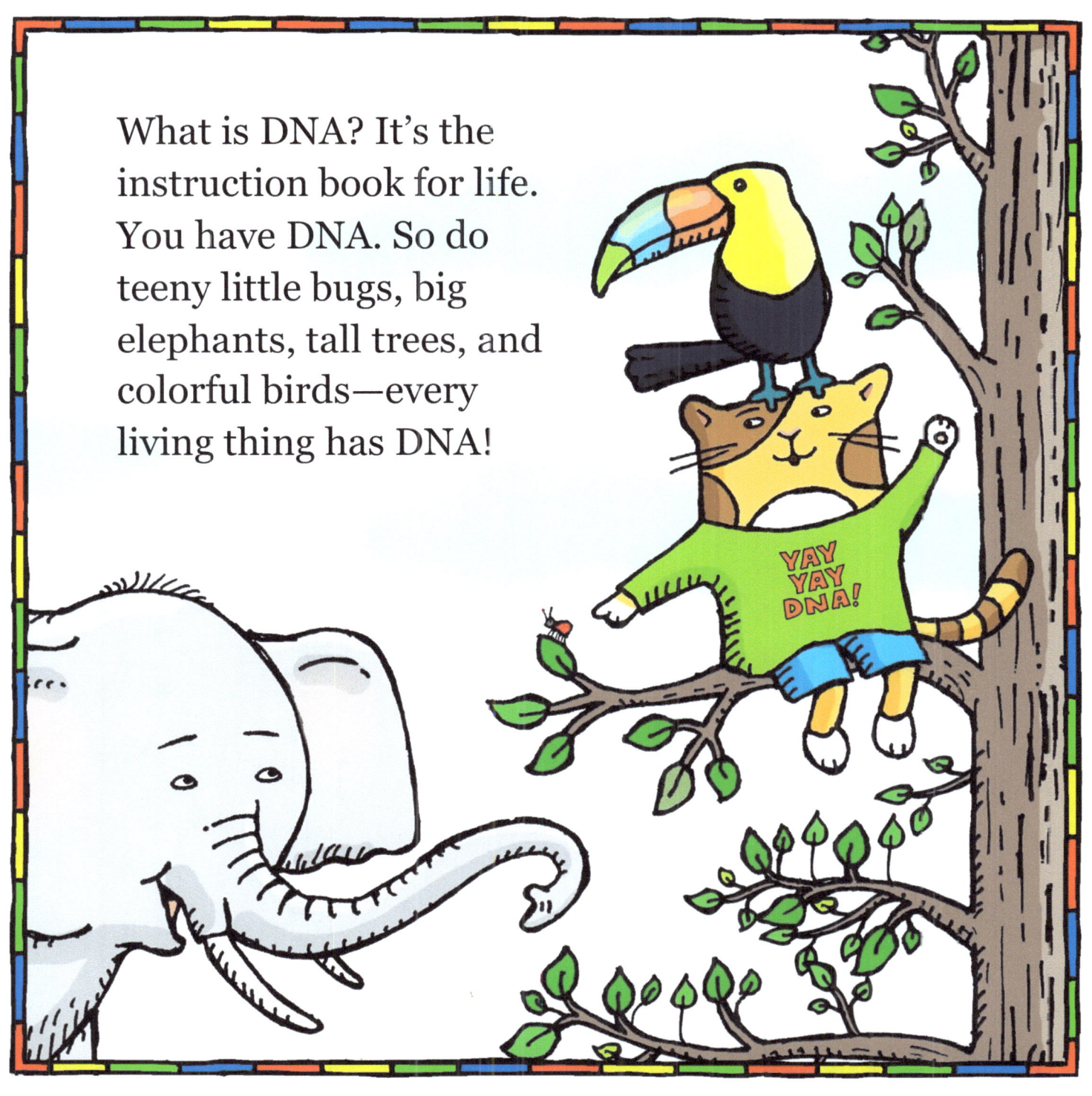

What is DNA? It's the instruction book for life. You have DNA. So do teeny little bugs, big elephants, tall trees, and colorful birds—every living thing has DNA!

You share about 90% of your DNA with house cats.

House cats share about 95% of their DNA with tigers.

And you also share DNA with just about every other living thing on Earth as well. A little more on that later in the book.

DNA is really small stuff. Your DNA is packed tightly into tiny threadlike things called chromosomes. Chromosomes are found in the center of your cells.

Your body is made up of cells. Lots and lots of them. You have over **30,000,000,000,000** (30 trillion) cells in your body! So that means you have a lot of DNA in your body.

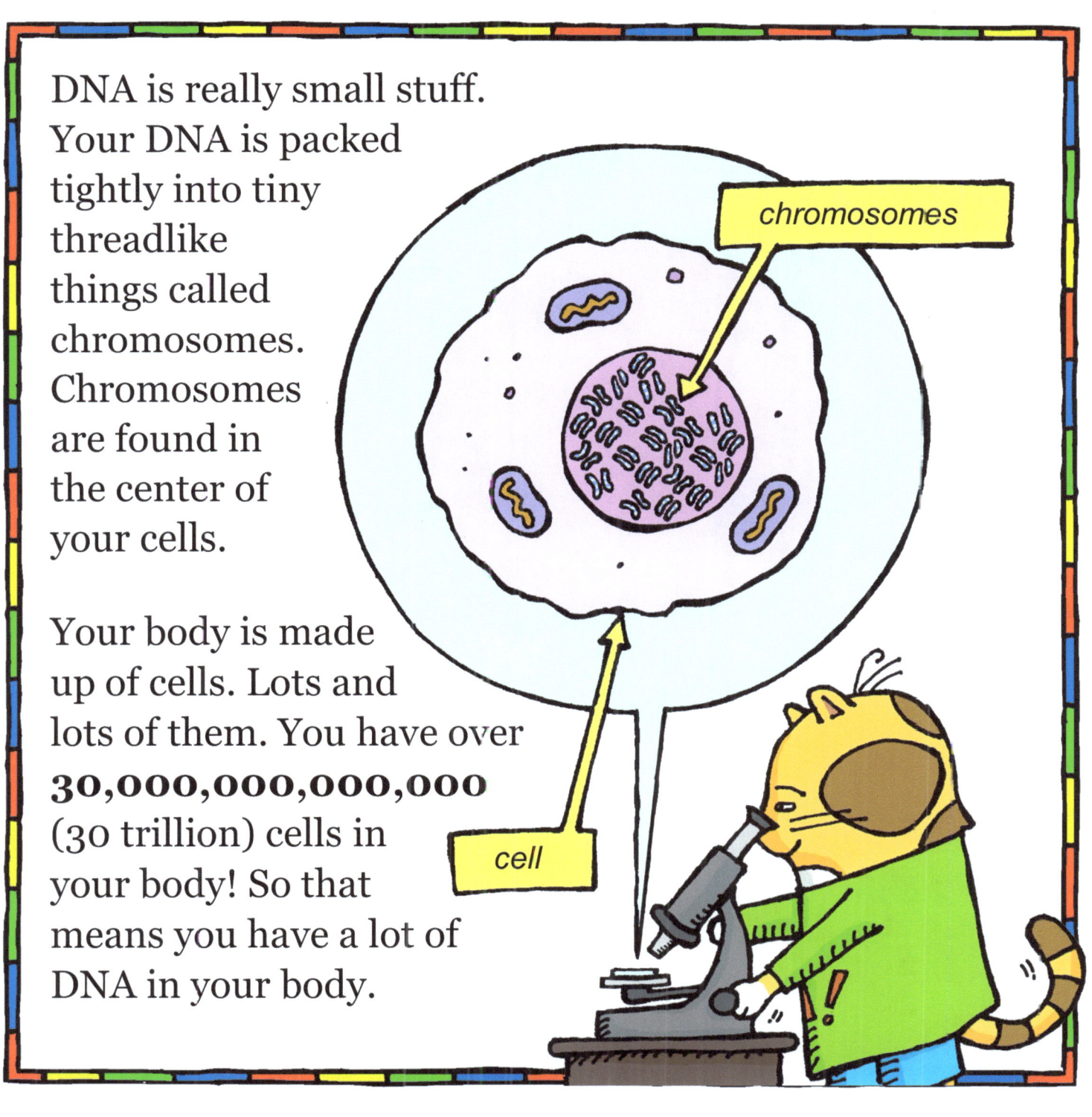

If you took all the DNA out of all your cells and tied it all together, it would be very long.

It would be so long that it would stretch far, far out in space. Way past all the outer planets in our solar system!

DNA kind of looks like a twisted ladder. And what looks like rungs on the ladder are actually base pairs of chemical compounds. These compounds are called nucleotides. DNA has four of them. There's *adenine, cytosine, guanine,* and *thymine*. To make things easier, the initials *A, C, G,* and *T* are usually used.

*DNA has an important little friend that helps it with all its tasks. It's called **RNA**, or **ribonucleic acid**. (rye-bow-new-clay-ick as-id) RNA kind of looks like DNA cut in half.*

There are several different types of RNA (mRNA, miRNA, rRNA, ctRNA, and others). All have different jobs to keep your body working like it should.

So, what does all your DNA and all those nucleotide base pairs do? Well, some of your DNA contains segments called genes.

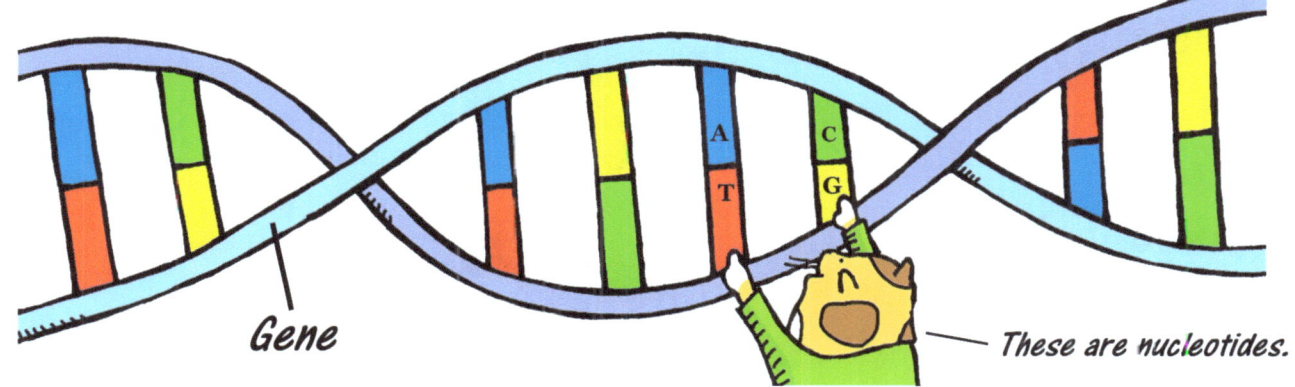

Every gene has a special job. The number of nucleotide base pairs and how they are arranged determines what each gene's job is.

You have more than 20,000 genes everywhere in your body, all doing many important jobs. Some genes are made up of thousands of nucleotide base pairs. That means there are billions of letters (A,C,G,T) of genetic code in almost every cell in your body.

Some genes help keep your body working. Other genes help prevent serious diseases. And some genes determine how your body grows and what you look like.

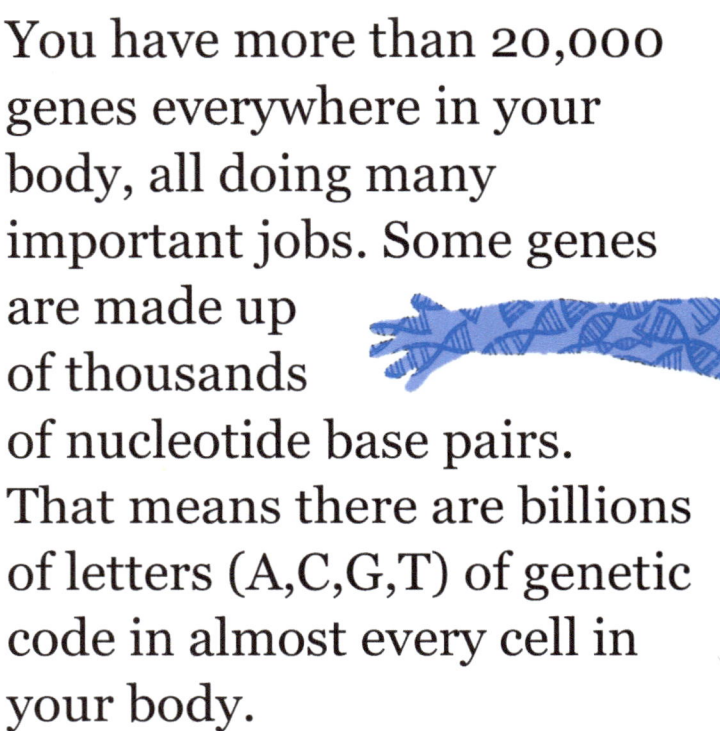

You have two copies of every gene. All your genes were passed on to you from your parents. Everyone gets one set of genes from their mom and the other set of genes from their dad.

Want to know something a lot people don't know?

Over 99% of the base pairs in your DNA are the same as every other human on this planet. You share a lot of the same DNA with everybody else.

However, there's a little bit of your DNA that is arranged differently from everyone else's DNA.

And that little bit makes a really big difference! That's what makes you you!

Do you have freckles? What color are your eyes? What color is your skin? Is your hair red, black, brown, or blonde? Is your hair curly or straight? Do you have dimples? This is all because the arrangement of the base pairs in that small bit of DNA affects how certain genes are expressed.

But you don't just share DNA with other people, you share DNA base pairs with just about all life on Earth.

Chimpanzees are one of your closest relatives in the animal world. You share 98.8% of the same DNA base pairs with them.

Over 50% of your DNA base pairs are the same as bananas. Go bananas!

Other life you share DNA base pairs with:*

Mice: 85%

Pigs: 85%

Chickens: 60%

Snails: 70%

Mushrooms: 50%

*Like all scientific information, these numbers might change up or down as new discoveries are made.

Believe it or not, all this DNA stuff started with peas. The kind you eat. In the 1850's, **Gregor Mendel**, an Austrian monk <u>and</u> scientist did experiments with thousands of green and yellow pea plants. His work helped launch the scientific study of how genes are passed on, and eventually the discovery of DNA.

But it took a lot of hard work by many different scientists over many years to learn what we now know about DNA.

Let there be peas everywhere!

Gregor is known as the *Father of Modern Genetics.*

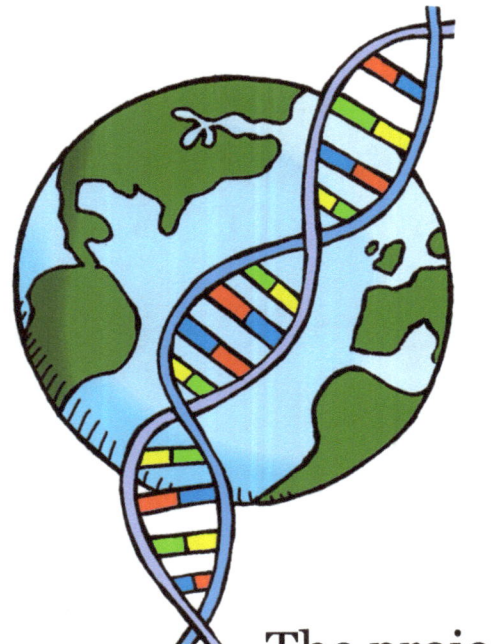

The **Human Genome Project** was one of the most important scientific projects ever done. It was a big scientific research effort involving scientists from all over the world. The goal was to discover what makes up human DNA.

The project began in 1990 and was completed in 2003. What was found has helped scientists to gain an incredible amount of knowledge about the human body, diseases, how genes are passed on, and much more. Yay Yay DNA! But there is still so much more to learn!

Some Words to Know

Traits
Traits are specific features or qualities that are inherited from parents. In people, they can be things like freckles, dimples, hair or eye color, or even something like musical or artistic talent.

Genetics
Genetics is the scientific study of how traits are passed on from parents to their children. It involves looking at how genes affect looks, traits, growth, and even the evolution of animal and plant life. Learning about genetics can also help scientists figure out how some diseases are passed down from one generation to the next.

Genomic Science
Genomic science is the study of the genetic makeup of different life forms. Genomic scientists study all of the DNA of living things as well as how the genes in the DNA work. Unlike genetics that may look at just a few genes, genomic science focuses on the entire DNA instruction book, or genome, of life.

Genealogy
Genealogy is the research of family history of present and past members of a family. It can help you learn where your family came from and a lot about their personal history. Genealogy uses documents like birth, marriage, death certificates, and other records to help create a family tree.

Family Tree
A family tree is a drawing or graph (sometimes called a pedigree) that uses genealogy information to show how family members, from the present to the long past, are related.

Genetic Genealogy
Genetic genealogy uses DNA testing and the study of genealogy records to help confirm family relationships. Genetic genealogy combines science and history to help gather more family tree information. It can also help discover important information about a family's health history, which can help prevent some illnesses.

About the author and illustrator:

Mark A. Hicks is a retired freelance artist and illustrator. He created thousands of works of art for children's books, magazines, greeting cards, and a whole lot more over the course of his long career. Now he creates artwork as a hereditary cancer awareness and prevention advocate.

When he has spare time, he likes to learn more about his family's history and his DNA. Using documents and DNA testing, he has been able to trace the roots of his family tree back thousands of years. And oh, what a family tree he has discovered! Some of his genetic relatives are quite historically notable. Mark has found that he inherited some interesting DNA from ancestors from around the world.

www.ingramcontent.com/pod-product-compliance
Lightning Source LLC
Chambersburg PA
CBHW061158030426
42337CB00002B/41